Pete Green
DEPOSITION

DEPOSITION

Pete Green

With drawings by Gaylord Schanilec

New Rivers Press 1982

Library of Congress Catalog Card Number: 82-61648
ISBN 0-89823044-6

Typesetting: Peregrine Cold Type
Keylining: Daren Sinsheimer
Book Design: C.W. Truesdale and Daren Sinsheimer

Some of these poems have appeared in THE NEW YORK HARBOR REVIEW and THE LIVING CHURCH. Our thanks to the editors of these publications for permission to reprint. Our thanks also to the United Arts Fund for its support of this and other projects.

New Rivers Press books are distributed by

Small Press Distribution, Inc.
1784 Shattuck Avenue
Berkeley, CA 94709

Bookslinger
330 East 9th Street
St. Paul, MN 55101

DEPOSITIONS has been manufactured in the United States of America for New Rivers Press, Inc. (C.W. Truesdale, editor/publisher), 1602 Selby Avenue, St. Paul, MN 55104 in a first edition of 1200 copies, of which 175 have been bound in boards and 25 handbound in boards.

For Max

(1891-1960)

Thanks

Deposition

3. Some Wonders

4. Some Perplexities

Opening Statement

With one long flinging prance
a year-old Boxer pup took the morning lawn,
upped and splayed his forelegs, toppled and rocked, tossing
his foolish buoyant bones

then snapped still
                                        though the drift of him carried
flailing and fifeing skyward,
coptering senselessly high as the lark—

So do I come
into this body of my homeland:
take up my stand, Earth,
here at your heart.

1. Fourteen People

Susan Waking

This round dawn sea
swelling under
my small boat
is the blue eye
of Earth.

Up from my warm
rocked sleep below
I wake—we
wake together,
the sea and I—

sure and still.
We stretch and gaze,
holding heaven.
I am the pupil
of this all.

Binky Cavanaugh

was mad
and sat under a pool table.
I know nothing else of the man
but that he got his soul through to death.
And here I am, a generation and a sea away,
on this Monday, at this desk,
remembering him.

Albert Keller

It is the angels' coming for him
that makes the most of Albert Keller,
his stubble, his shuffle, his days' end.

First he fell in the snow; now
in the grass an inch too deep. *Atsa*
stuff, he breathes, when I pull him up.

They come   swinging down from the sun
like two great kites of glass
ablast with the face of the sun as mirrors

out of the blue tree of the sky

Mr. Keller stands on his sidewalk
staring at the keys coming unsorted
in the palm of his hand   reeling them in.

The Comte de B. & Bernadette

One day in 1858
the Comte de B., from summering
nearby, dropped in at Lourdes to air
a civil doubt or two. Arrived,
advised, he made directly for
the subject's house, unfurling as he went
his graciousness. Bernadette
sat on the front step patching socks.

He questioned her kindly;
and thoughtfully, intricately,
thoroughly. At last she said,
You think I'm a liar.

No. My dear, these dreams are like
your wooden shoes: they're yours, they suit you.
But they'd cramp me, and others, awfully.
You see, your figurings, your tales,
even your very person,
are an attempt at definition:
And, child, all our definitions
cramp the great Indefinites.

She got up. Because you are a sinner
I will repeat her smile for you.
She pressed the palms of her hands together
and began.

Why did I come? he wondered *What am I doing here*
so far from the convenience
of my being   me?
This bone's edge jagging through
the skin of things   is not my business.

Then, the way the story is told,
he clumped down there on the road
in the face of his seeing.

Richard M . Nixon

You may own the forest
and the sawmill and thank
the God of wood for both
and still run out of plank.

Little One

Little one, we were not ready for you
when the sheer news came that you were coming.

None of us was. You were coming anyway
sure and slow as the false dawn lifts

over the humming plain. Then you weren't, and again
we were not ready as the light collapsed.

Now, years later
I sit alone in this drumming city night.
We miss you. Wait for us. We're coming.

A Homemaker's One Monday Morning

God I felt awful.
Blood like cold tar.
Feet 2 cinder blocks
ready for the East River.
Kitchen thick with kids
finishing breakfast, fending off school

then, loud as dropping a drum,
Mike, ya ready?
Benjy Rosenthal at the door.
God just one more...

But there he was:
lazy little bastard Benjy
with our empty garbage can in his arms
lugged the length of the driveway
all on his own,
grinning.

And the day flipped over
easy as a pancake,
crisp,
ready to do.

Preacher Preacher

Preacher preacher come to call
in his shoes and smiling
sees the mud, his duty too.
In his socks still smiling.

How are we and do we know
Jesus Lord all smiling?
Yes, in fact, we do. But,
well, not always smiling.

Know him quick, abrupt, dark,
and know beyond his smile
his heave, his curse, his roar, his howl,
his galling, even guiling.

Yes smiles preacher yes I know:
In fact
though my feet are right now cold
still it leaves me smiling.

Celibate

(Speaking to one not the Lord Christ)

You are become
my daily bread
now I know
you are you,

bulking, sopping
the day's light
the air
by only your being,
or having been, here.

Now I know
and now I must turn
and set me to
what will taste
and chafe like death:

forfeit bread
light, air
and ream the day down
to its old sane core.

Mouth to mouth
we're flesh and cross
cross and flesh
but borne, alive.

Drawing the nails
will tear
as, into another,
driving them.

Young Mother

Stepping in I see only
the headtop of fine black hair,
the blue dress, not
the steel rake dragging
the roots of her womb.

Four days ago she gave forth
her first child, a-wonder
that the months had been heaping to
this very one. Today
at first light he died.

The tears hit and print
a blue almost black
in the sun on her lap.
My words comb through
only her shining hair.

Mary

Where ya gonna put him, Mary,
now he's here?
Oh, I'll just hold him
till he gets down.

What'll ya do now, Mary?
Look, he's down.
Oh, I'll watch him play
till he goes out.

What'll ya do now, Mary?
Look, he's out.
Oh, I'll wait, you know,
till he's all grown.

What'll ya do now, Mary?
Look, he's grown.
I'll—oh look, how lovely,
the soft-falling snow—

I'll go on cooking, cleaning,
till he's gotta   you know.

Mrs. Peña

Full of the town I come again
where she sits with her knees drawn up
and her head down
and the flesh of the wall of her neck
blown through from the inside,
blood and pus oozing.

She sits. Before God and all.
Leaning into the mystery.
Her reverence more than a town.

Etienne Gilson, Historian

1.

Night. He goes down into the pond-bottom-dark
and crouches among the caved-in hundreds-deep shelves of the
leaves.
Chooses one,
a fragment of leaf:
places pencil tip to a point along the leaf's edge and begins,
tip by tip, the month's-work tracing of one sub-vein's mole-way
back,
twig and branch, through drifting liquid dust,
towards the great spine-vein's slowly flaring

gone

stops.

Presses finger- and thumb-pads into the flesh of his eyelids,
kneads.
Takes pencil,
jots
on a fresh white card going sodden under the lead.
Takes paper,
writes
in the small moon of the lamplight on his table

writes

under
the hovering
fantastic porous phantom
of the night

of the past.

Writes.

2. (*Some of his readers*)

Phhht a can opens, years, a sea, away.
Aggie reaches across the light, grinning.
Will snorts above my 7th flourished poet.
OK gang, let's not, Grimm begins.
A goose-neck student lamp tilts in the rubble.

The night lodged in their throats and hearts is logs, is longing;
the stuff of it up-bulging, pressing like the thumbs of hands
bent on getting out; brawling against concavities
of skull and eyeball; ubbing live and blind as sap;
convulsing in kettles over the wind-spiked fires.

Like that Irishman in the Gospel! one guffaws.
Another plays the body of his voice like a trombone
(like a driver alone on his bus parked in the rain,
up and down the aisle, strutting and pumping),

while the 3 crows of order, clarity, meaning,
perch on the limb of the night's horizon,
on the curtain rod,
flare Awp!
flap off—

collapsing into themselves,
subsiding.

3.

Writes

out of the many-ponded sea of the past,
of the night;

out of his imagining mind spreading denser than the sea
(as the sea is denser than the night),
huger, rarer than the night
(as the night is than the sea),

drawing its net
infinitely slowly

up, in,

with enormous pondering care.

We, all, compassed in his words, his love.

2. One Man

At 39

Just now I know
the heart's a vise
whose jaws begin
finally to move.

I'm walking back
from dumping garbage,
up a grass path
to the house, miserable,

as it begins
with one first
slow fibrous
tear, opening.

At 44

I pitch through a hole in me.
Black. Nothing moves.

Way up, a manhole
like a dime of air

slowly shows
no pictures, no walls.

Maybe nothing has ever moved.
The dime winks and goes black.

Near Middlefield, N.Y.

The tall corn alters the road it stands along,
taking the barns down into itself, blowing
over the sheds. A car half tunnels along.
It's all new. It's old. An aisle of the first shrine.

I may know where I am
but am I still the one
sent out in my name
those years ago?

Earth, Thou Beauty's Smouldering

Earth, thou Beauty's smouldering,
like fire in a pile of sticks
winking down at the edge of a road
on an Autumn evening:

I squat, poke a coal loose,
scorch a finger, damn,
finally get a cigar lit
again and again and again,

blind for the while to even you,
and to the waking dusk beyond.

At 46

1.

You say it is unnatural that
your pear tree's branches, hung with only
the weight of the pears themselves, even
granting a full splendid crop, should crack
and splinter and hang helplessly so
over the golden sprawl of the pears
as you found them the other morning.

2.

Here's a second case:
squat trunk rooted like every other
in packed-dirt dark, bearing faults up thin
as batsqueak, and bearing their lacing
echoes back clear in to heartwood—
twisting, swagging down almost to earth
from only the weight of my life along its branches.

Oh, there's more:

the light and the wind,
the high blue days
rounding and folding
into night,

and imagine holding
Jenny's eyes
over that crowded
Bakery counter,

like upfalling fruit,
the pears of dream,
or a white-gold stretch

of the sea at noon,

a tern's flickering
skim and plunge
in the sun of that sea;

beneath, my Jean,
a sounding whale
all black and white,
frolics and flails,
among the roots of the sea.

But these weigh too
even as they feed
and any talk of gathering's
years away still.

3.

Considering closely now,
whose life do you know
that is not overborne?

A Quiet Day, Reading

The sun bumps gently down
behind a small bare tree
closing round trunk and branch
like fingers round a wrist.
The tree sinks into the sun.

I take this coin, this dole,
like the head of Thomas More,
like Peter's sprint for the tomb,
in love and loss and longing for
the blazing face of the whole.

A Few Steps Into Joyce's *Ulysses*

This writing man's inside is damp.
It drips and bores like a cave,
winding down and round
to a last black snarl.

Of course, you
may be fond
of caves, and go at them for science
or fun. Me,
I'll take an old-fashioned shore and the sun
snapping like castanets on the sea.

To be fair now, from the outside
a cave's not so bad:
it's even a little promising;
intriguing; a mouth; Earth
might spell her insides out
from here, if she'd a mind to, we say.

Conundrum.

I'm no guide in that dark
our living is bent upon

that utterest

what with my tastes and all, and being,
as I am, grandson of a Murphy and a Doyle.

To His Own Soul

The two young men of our office had quarreled
shrilly. Then at five o'clock
one pranced grandly out
and down the fifty steps or so
it was to the corner: in that space
he looked back fully twenty times,
jerking his elaborate blond head
dizzily around again and again
with the bursting back, the revolutions, of hope.

A unique exhibition, I guessed,
to be displayed at last
some place like this.

But now, now, now

so do I go on
seeking you,
spinning and gaping,
groping the crowds
and hours for you
(you who are already here by my side),
seeking you
with the brightening, blanking face of hope,
wantonly, insisting like the sun.

On The Road Above Casiguran

1.

One night more than twenty years ago
in the one lit room of a big dark house
a girl bent over a basin slowly washing
herself. I stood behind and gently,
solemnly, moved my hands upon her
flesh   could not be more soft and still be sure.

2.

Dust, blaring sun, this mountain logging road
Bang   goes the parish bell. That's for Alejo.
They brought him in from the barrio yesterday
his back a blaze of machete wounds.
Bang   blaring sun, dust, this whipping road
taken too fast, springs gone on the jeep,
winding, unwinding, the endless rail of the wheel.

3.

Then I took a tithe of the mystery. Now
the nine parts come on at me and I writhe.

Leaving For The City

It's hot as inside an old bus stuck in the sun
windows jammed shut
at that blank space around noon.

A door opens
out step two old Jews
black long coats, black hats
white beards in the sun

they walk to me
and pass
on this day of my leaving for the city.

Muffling the sound with a fist
I cough twice in the silence
of my heart, clearing my heart.

The silence holds, then breaks
as the movement begins, the last
I guess, here
on a corner of a summer noon in Cobleskill.

A Long Look In The Mirror

Let's face it, Green,
that's you-in-the-face
at 47:

bulge of skull,
eyes behind glasses
skewing just off true,

that stretch of mouth
(nest of words
and silences,

haunt and launch
of extravagant birds),
that yank of beard:

wanting the world
under you
like a saucer
under a cup,

and you safe inside
as a half-swallow
of cold coffee;

furtive as a shadow
at last light.

Flesh to mirror-glass
crying,
I do not know
how to live;

mirror-glass to flesh
replying,
Indeed,
indeed.

That's it:
(except for what
no mirror knows:

the counter-extravagance of Grace,
cunning as a thumb
opposing fingers),

the cards, your hand,
play.

East River Christmas Eve

Like a barge on the river night
deep and still he comes
under, past, unsearching lights
from Queens. The tug thrums.

I lean and wait

longing to lurch up and empty me
like a barge, or a bell: out
with it all, pitching wildly free
in one boundless shout!

Lament Of The Cultural Attaché

(Poet approaches Public)

Misjudging the line's stately undulations, pace,
(oh will they live me in some far, far land?)
I'm checking my zipper as
the Queen extends her hand.

Everything's About The Same These Days

1.

but these days are different. It's not
exactly the look of things; nothing so strange
as that greening light coming before
those summer storms when I was a boy.

Using the old image of journey,
ascent, I wonder is it only being
over the top of my hill of time, being
on the far, the down, slope?

2.

Dawn.
The sun
brushes
along the bricks
of crosstown house fronts—

Later
it palms
the new-washed
heads of hair
bobbing, cocking, turning—

It combs
and combs
the great curling back of the bay:

Incipience.
The smithereens
of glory.

3.

And here I am
at 48,

again
about to begin.

3. Some Wonders

The Moon Rose Round As A Button

The moon rose round as a button.
The tree blew soft as a sigh.
Under both, near love-lock,
lay you and I.

One breast-like cloud swelled gently,
spread, rolled, like a thigh.
Deep in summer shadow
lay you and I.

I'll love thee till the moon works loose
and hangs by a thread from her sky.
Till I have cast my breath beyond
wording her last cry.

You Are Light To Me, I'd Lied

You are more
than light to me.
You are you.

The eye's share
of the sea is
one slice, one

layer of blue.
The blind heart drinks
the full sea.

So Shy And So Sure

So shy and so sure
my love comes to me,
comes like the blue
of dawn to the sea.

So too
comes the Mystery,
the long sembling
shock of the Trinity:

the old one at table
lifting to love
wine bright as the bride
on his son's knee,

and Fra Angelico
dancing with Eve
while Dr. Johnson
listens. We

exult in you,
God, that you are
lovers like us
and wine and thirsty,

for coming (oh
come, thou Kingdom,)
as Jean comes to me,
like the blue of dawn
to the spring of the sea.

For M.

The bridge under that bus you might have been on
went down in rain and wind into the river.
I see the river tumble and sog your body
taking it down to the sea,
and out, and on,
though I know you are near and safe enough.

Today the sun is bright on the quiet surf.
When it comes to bodies
who will ever be
near and safe enough?

On A Road Map Of New York

On a road map of New York
you could cover with a quarter
the three places she has lived
since she was born

who now gladdens all the seven seas
and whole wide countries
with their houses, down to a room,
a chair rocking,

of this old world
of my heart.

B. Likened To The Moon

Some days ago you were
one roundish white chip of cloud
among the bits and chips of cloud
against the blue there.

Now you take face and color
and become the precious one
for whom the whole grand process
of sky and star proceeds.

I cannot say how it was
once that heaven without you was enough—
that boarded lot of silences,
that blinding rear of gold—

and yet it was. And yet it will be.

Old Fields Of January

Old fields of January
along Route 15
lit from within like a sky
the sun is about to step into;

broom-sage standing thinly
in fields of weathered tan,
itself the color of sunlight
standing in beer, and moving:

these are colors and reaches
of a gleaning Love,
and they bare the grazing
body of our love

in the blue, delighting air.

Love moves: and these fields, and the whole
undivided body of Earth
moves, colors and brims,

as I move, and my Jean
moves under me, and we move
together in Love's delight.

Three War Poems

There Is This War

1.

There is this war
and since I'm here
okay I'll fight
and maybe die.

But who are you
and what is death
or even life,
my brother?

Me I'm one
of Adam's boys
and life is keeping
your blood at home

where she belongs,
in off the streets,
out of alleys,
ditches, fields.

Death is when you don't.

2.

Blind as bugles
our nights call out
to you, O Lord,
heart of morning:

night and morning
throat and heart
endlessly
exchanging calls.

3.

Lift his body
up for keeping
from the mud's suck
when he

Catch his blood
in a gold cup

when he falls.

That Land

What can that land look like, smell like, now?
Look: Hang a man by his hands. Sink
a branding-iron into his lower belly.
Now hold it. There. There's your land:
the stink, the smoking flesh, the smoking hair.

Christmas, 1981

Polish blood steams in the Polish snow.

That ancient virgin bride called Longing,
called We-Will-Be-Free,

clubbed and down,
bleeds.

And see:
in the numbing wind,

the frozen hollow night,
again she thickens, gives flesh.

Freedom? The Child? Only and always
out of the belly of this holy one

this holy thing comes, and, for the time, bides.

Christ Hoarse

I am your word
gone skimming and slithering out across
the wet streets of the air
wrenching awry
the gaze of your people.
I am a needle passing through the camel's eye.

Christ Cornered

Dim, dimming,
down to 9 Watt,
loose in your socket,
webbed over with dreams,

you

wrought with such cunning,
set with such care
to light the world.

Christ Down

1.

These olive trees
keep out the moon,
keep in the gentle
snores of my friends.

2.

I have built no house.
I have stretched no field.
I have got no book,
no child. I am alone.

Beyond even the Christ-threat.
I've set no stone upon stone.

Deposition

(Christ is speaking from the cross)

Young man, you down there
plotting that magnificent Deposition
cloudy with sky-
hooks and pulleys

I am tired. Get me down.

You do not warm me in even one member
with your endlessly figuring eyes
your jottings
your chafing gape

I am numberless. I am cooling.

Let me down. Look
I am only this broached heart
this armload or so of bones.

4. Some Perplexities

Summer

Way out a ship goes by
slow as the day almost.
Children fret the sand.
Under such
a sun
the sea shuffles in place.

Faith

gropes in
shocks itself

on the charged barbs
of the Law

lurches on
in shreds
in raw need

(the Law coiling
and writhing behind)

rends
the temple curtain
the virgin's skirt

shudders
blackly
on

Fat Slow Bears

Fat slow bears,
wide upright elderly amiable creatures
with maybe a fresh Blackeyed Susan
for buttonhole and bulls-eye:
if you must be trying us
next time send bears.

Today there was a copperhead
sprawled across the yard
just below the back step,
bronze and orange and long.
We stepped out almost on him.
Shock sucked us back like a breath.

I ran for the gun, clambered
onto the roof and blatted
down eight rounds. Dirt
jerked up round its head.
The whole forebody lifted, probing,
then dipped and slid on under the house.

Jean stood at the window very still.
We've been here three months now.
This house has leaky floors.
Already something, ghostweed, the shimmer
laid down by olden moons, washes up
in and out of this house of our love.

This Poem's About An Old Love Gone

This poem's about an old love gone,
a new one not quite come,
and it's sad as the whole sea.

The one stands raw with tears at that sea's edge
as I row a small boat out alone.

Between me and the other—
me shuffling here in sand,
she moving restless there on the far shore—
that same sea swirls in the glass of itself.

I cannot reach and hold her till
I've drunk it down.

As A Seed Is

What went on between us tonight
after ten years
and the three last terrible days

was small and dry
as a seed is,
and maybe grander than

the apparentest standing marriage
flaring glossy as a Magnolia
in the pouring southern sun.

At Noon You Went Down

into the subway at 77th street,
turning at the bottom step,
raising and cocking a hand,
lifting a smile.

I stalk west on 77th,
restless, leaning into myself,
taking my new coat for a walk,
on the first cold day.

I watch the children climb
the Alice statue,
hear the parents' old languages,
stare into the drained pond.

6 hours later you haven't reached your place.
You aren't there.
The phone-rings strike
out and out.

Like casting an unweighted line.
Every 10 minutes I try.

The lake of the night is empty
of you.

A 4th time, a 5th
Hello

you say and my heart jerks up
like a net
tightening all round,
biting in.

Cobleskill

The evening star
deep behind
a white winter
village church,

rose and snow
poised about
like a set
of ground-up angels,

is not enough.

But there it is.

I wish it were.

On Parting

If I cast among
none after you
I shall go lonely
and bitter. But

if Jack come and Jill,
even thronging, quick-
new as scene-change
or city, yet

each wanting so
soul and tongue I'll
still long for you.

Nights

Old jazz struts and snakes out of the radio at midnight
from the other room where Jean is reading ancient tales
under a hanging lamp among the shadows and the new furni-
ture:
it's red and yellow and black and raucous and still,
a cave
cut into the heart
of the immense mountain of this country
night.

The brook sounds.
A cornet flares bluely, fades.
The mountain breathes in the night of God.

I doze. We're stage refugees, pushing a cart, pulling a goat,
backpacking a parent each; children, cats, lagging, clinging;
*Please Benny chust for my sake! Where did I go wrong? Mom-
ma*
he's pinching me again! mattress, rocker, boxes, papers, tools,
doubts, dreams piled perilously high:

going
hand in hand
our shadows and ourselves
into the only night.

Spiritual Prose

God, how they go on
about your *approach*,
its well-bred *delicacy*,
restraint, *exquisite grace*.
You are a bright wet fish swung
by the tail, splat, in my face.

And how *tenderly*
you work in them,
as breezes gently shape and swell
silken summer jibs.
You are comfortable in me as
a set of the wrong size ribs.

Untitled

1.

I will kill you pounced out
of my breath into her whurling jaws

and my arms shudder awake
rocking and black-fisted
under the roaring round:
hammer-bones cock
click their flickering, splitting
tongues murmur

We are ready. Heart
blurts *I am not.*

Ears gape like boys.

2.

Yesterday
we sat in a restaurant on York
eating tuna fish sandwiches
fat as old bicycle tires,
laughing: remembering
the waitress who'd bobbled the service
our first day in the city:
Is it your first day too? I'd asked.
The waitress had smiled assuringly,
I hav bean wi sees companee seventeen years.

3.

Tonight you say
*Don't get run over
with that poem on you.*

Reading The Gospels

1.

Finally the raked pages
are one page.
One slice of flesh
that, lifted,
looks through to a blank, black sea
where from miles down
one blurted word
glistens up
breaks into air
cries itself out

leaving the taut ear and heart forsaken.

2.

There's a bum with a souring know-it-all face
who comes each night and sits on the sidewalk
of 39th near Madison. He drapes a gray blanket
over his head and wears no shirt.
Sometimes he munches chunks of bread
from the pavement. Mostly he stares.
He's worked-in there, in the way of this city's towns,
by the shopfolk and cops and striders-by:
Francis without the first degree burns.
Who doesn't read the gospels
and so keeps clear of the cry
would take that front off his face
the way a machete takes
the end off a coconut.

This Business Of Making A Poem

is feeding lines out into the current,
day-long playing them out with ridiculous care,
(no bait, no hook, are the rules),

the water silking along their endless sides,
the lines tracing exactly the run and roll
of the days,

the feel, the heft, of the thing
just this side of nothing at all.

Staten Islanders' Homegoing

1.

Tired, we troop on and on to the ferry
while it's still a chunk of the land   snugged
by two short cables with hooks bigger

than hands   till a whistle *Krees!*   and the stepping thins,
stops. Gates clatter. A pin whams
home. Un-grappled

on the water,   till at last
rocking slightly, then push with a fat-sided whooshing
out into the channel

swinging smartly to the right,
steadying,
then slowly left.

The moon,
full as only the moon is, swells
over the interlacing length of the three bridges

and the bay opens out at the feet of the city.

2.

Dark themselves, the buildings go up
in rungs of light.

The bay broadens,
like sleep.

3.

The ferry clambers on
through the knuckling water
tumbling time with un-time,
kneading darkly along.
Light-trace of tower,
buoy, tug, barge,
plops and wobbles like dice
on the water, till at last

4.

the city glows at the rim of the bay,
its light making intricate sense
of its dark;

all its bulk becomes one precious thing;
like the secret at the core of your most gracious dream;
or waking.